Kali Lin

Kali Linux Made Easy For Beginners And Intermediates

Step By Step With Hands On Projects

(Including Hacking and Cybersecurity Basics with Kali Linux)

Craig Berg

Copyright © 2020 Craig Berg.

This book or any portion thereof may not be reproduced or used in any manner whatsoever without the express written permission of the publisher except for the use of brief quotations in a book review.

Introduction

Kali Linux is a Debian-based Linux distribution designed for penetration testing and forensics. Kali is ideal for penetration testing because it has in-built tools and utilities that make this task easier. It has over 600 pre-installed tools each of which aids the tasks of penetration testing and forensics.

First introduced in 2012 as a successor to its previous version Back Track Linux, Kali Linux runs as an operating system that you can install on a hard disk, run virtually, or live—the first part of this guide covers Kali Linux installation. Offensive Security, a firm whose core developers are Devon Kearns, Matt Aharoni, and Raphael Hertzog, maintains Kali Linux.

NOTE: Although Kali Linux has some of the best tools and utilities for penetration testing in the world, for this hands-on, Kali Linux guide, we may have to install some other tools needed to make it easier to learn how to use Kali Linux for penetration testing and cyber security forensics.

The following are some of the major features of Kali Linux:

- ❖ It is a Debian-based Linux distro.
- ❖ It comes prepackaged with over 600 tools and utilities.

- It supports ARM architectures.
- It has custom kernel for packet injection.
- It is an open source distribution.
- It is Filesystem Hierarchy Standard Complaint.
- It has Multiple Language Support.
- It is fully customizable per user needs and kernel rebuild
- Defaulted with gnome desktop environment
- It has advanced wireless card support.

In this book, we are going to cover the major features & tools provided by Kali Linux. They include:

- Information Gathering Tools
- Vulnerability Assessment
- Wireless Attacks
- Web Application attacks
- Exploitation Tools
- Forensics tools
- Sniffing and spoofing

- ❖ Password cracking
- ❖ Maintaining Access
- ❖ Social Engineering tools
- ❖ Reverse Engineering tools
- ❖ Hardware Hacking tools
- ❖ Reporting tools
- ❖ Denial of Service Attacks

We shall cover each of these features & tools individually so that after reading this guide, you have hands-on experience with using Kali Linux and can use what you learn when completing the hands-on Kali Linux practice project found in the part 17 of this guide.

PS: Don't forget to leave a review of this book on Amazon if you like it!

Table of Content

Introduction ———————————————— 3

Part 1: Installation and Configuration ——— 12

   How to Do a Kali Linux Hard Disk Installation   13

   How to Do a Kali Linux Virtual Machine Installation ———————————————— 16

   How to Run Kali Linux in Live Mode ———— 20

   Installing Vulnerable Machine ——————— 20

   Updating Kali Linux ———————————— 22

Part 2: A Basic Introduction to Networking and Working with the Linux Kernel System ———————————————————— 25

   Linux Directory Structure ————————— 25

   The Linux Command-Line ————————— 28

   Networking Basics ———————————— 34

   Network Hosts —————————————— 35

Network Protocols & Services 35

Network Ports 38

Part 3: Information Gathering Tools _____ 40

NMAP & Zenmap 40

Netdiscover 43

DNS Tools 43

HPing3 44

Part 4: Vulnerability Assessment _____ 46

Web Application Scanners 46

Network Vulnerability Scanners 47

Part 5: Wireless Attacking Tools _____ 50

Kismet 50

Wifite 51

Fern Wifi Cracker 51

Part 6: Web Application Attacks _____ 53

Database Tools _____ 53

CMS Tools _____ 55

Part 7: Exploitation Tools _____ 57

Metasploit _____ 57

Armitage _____ 61

Browser Exploitation Framework (BeEF) ___ 62

Exploit Database _____ 64

Part 8: Forensics Tools _____ 65

Autopsy _____ 65

Binwalk _____ 66

Rkhunter & Chkrootkit _____ 67

Volatility _____ 67

Hashdeep _____ 68

DDRescue _____ 68

Part 9: Sniffing and Spoofing Tools _____ 69

ARP Protocol 69

Ettercap 71

Wireshark 72

SSL Strip 75

Part 10: Password Cracking 76

John the Ripper 76

Medusa 77

Crunch Password Generator 78

Rainbow Crack 79

Johnny 80

Part 11: Post Exploitation 81

Weevely 81

Powersploit 82

Shellter 83

Nishang 84

Part 12: Social Engineering — 85

Social Engineering Toolkit (SET) — 85

U3-pwn — 87

Ghost-phisher — 87

Part 13: Reverse Engineering Tools — 89

Dex2jar — 89

Java Snoop — 90

OllyDbg — 90

EDB-Debugger — 92

Apktool — 92

Part 14: Hardware Hacking — 94

Arduino IDE — 94

Part 15: Reporting Tools — 95

Dradis — 95

Magic Tree — 97

Part 16: Denial of Service Attacks _____ 98

Understanding DOS _____ 98

Tools used in Denial of Service Attacks _____ 100

Xerxes _____ 100

Part 17: Hands-on Kali Linux Projects: Exploiting Metasploitable 2 _____ 102

Requirements _____ 103

Network Scanning – Nmap _____ 103

Exploiting FTP _____ 104

Exploiting SSH _____ 105

SSH RSA Exploitation _____ 106

Telnet Exploitation _____ 107

Java Exploitation _____ 108

Virtual Network Computing _____ 109

PHP (CGI) _____ 109

Conclusion _____ 111

Part 1: Installation and Configuration

In this section of the guide, we are going to look at how to install and configure Kali Linux. We will cover both hard disk installation (dual boot) and virtual machine installation.

Before we can install Kali Linux, the first thing we need to do is download the ISO. To do so, go to the following web page:

https://www.kali.org/downloads/

On the download page, you will find various download options based on architecture and Kali image type. For this tutorial, we are going to start by installing Kali on the hard drive; to do this, download the ISO version as highlighted below.

https://imghostr.com/60d818_ha7

You can select the default Kali ISO (marked above) that uses the gnome desktop environment. However, if you want to use other desktop environments, you can download the KDE, XFCE, MATE, and LXDE.

NOTE: Do not download the light version of Kali Linux because it does not have all the defaulted tools and to use any of them, you will have to install each tool individually, a tedious process.

Once you have obtained a copy of the Kali Linux disk image, you need to compare the SHA1 hash of the image you just downloaded with the one provided on the download page. This ensures the image you have acquired is untampered and incorrupt. If you are running Windows, you can use the hashMyFile tool to check for the sha1 hash validity. You can download the tool from the resource page below:

https://www.nirsoft.net/utils/hash_my_files.html

How to Do a Kali Linux Hard Disk Installation

The most convenient way to use Kali Linux long-term is by installing it on your hard drive. You can install Kali Linux alongside other operating systems such as Windows, MacOS, and other Linux distributions.

To install Kali Linux on the hard drive, we are going to require several tools.

They include:

- USB/DVD bootable media
- A minimum 20GB disk space for Kali Linux installation
- Kali Linux ISO – obtained above

Installation of Kali Linux on a physical machine requires a clean, non-partitioned hard drive without any data in it. Here, we are installing Kali alongside another operating system.

Once you have obtained a media device such as USB or DVD, it is time to create a bootable disk. If you are using a USB flash drive, download and install BalenaEtcher tool and follow instructions to burn the image. You can download the tool from the resource page below:

https://www.balena.io/etcher/

If you are going to use a DVD, download and install Burnaware:

http://www.burnaware.com/

Once you have created your bootable media, you need to create a partition for Linux partition. If you are using windows, open disk manager, select the disk to partition, and click shrink volume. Set the size of the partition—it must be 10 GB and above—and click shrink.

https://imghostr.com/4c7c48_dpi

Once you have allocated your partition size, boot your device using the created bootable media above.

NOTE: Different computer models have different boot configuration. You can search the internet for your device boot options.

Once your system boots, Kali Linux will prompt you with the following boot menu window.

https://imghostr.com/9d4oce_m8o

Choose the method of installation. The graphical installer is the friendlier to beginners. Select and configure your preferred language of installation. Next, select your geographical location. The next step is to configure your keyboard settings. Follow this with setting up your network settings and connect to a network. After that, create a password for the root user, which is a MUST-do. Continue to setup user account and password.

On the next screen, which is the partition part, select 'use the largest continuous disk space' and select the partition you created earlier. Confirm and allocate the disk space, and then select 'write changes to disk.' Select 'yes' and click continue. Wait until the installation process completes.

The next step is the package manager. Select 'yes' and click continue. If you select yes, make sure your device has an active network connection. If not, select no.

Next, install the GRUB bootloader on to the disk. Once the installation has completed, reboot the computer and select Kali Linux in the GRUB menu.

How to Do a Kali Linux Virtual Machine Installation

The other way to use Kali Linux without installing it on the physical machine is running it as a virtual machine. Using virtualization software, you can run Kali Linux on your current operating system.

To do this:

The first step is to download a virtualization software and install it on your computer. In this guidebook, we are going to use VirtualBox from Oracle available from the resource page below:

https://www.virtualbox.org/wiki/Downloads

NOTE: You can also use VMware Workstation but this guide does not cover the VMware configuration process.

Once you downloaded and install Virtual Box, go to Kali download page and download the Kali Linux VirtualBox image. Once you have the image file, extract the file and save it in a folder. The extracted file contains a .ova file. Open it and select VirtualBox.

The next window allows you to customize the location, memory and other configuration of the system. Once you are all set, click import to start the process that imports the operating system. Once you complete that, click on start to boot the new system. The default username is **root** and password: **toor.**

The steps are very similar if you are using VMware version.

Virtual Machine Configurations

After creating and launching Kali Linux using VirtualBox, we need to set some important configuration so that we can completely setup a penetration-testing environment.

Here are the most important configurations:

1: VirtualBox Guest Additions

VirtualBox guest additions are 'required-installation' once you have launched the system.

They provide extra system features such as:

- ❖ It enables the system to support full screen mode.

- ❖ It enables host and guest systems to share files and folders

- ❖ It allows copy and paste between host and guest machines

- ❖ It Fixes errors that may occur with the mouse between host and guest systems

To install VirtualBox guest additions, go to Devices tab on VirtualBox menu and select install Guest additions. If you receive a warning message, ignore it and click cancel to continue. Next, open a terminal window and navigate to guest additions CDROM mount point as follows.

Once in the directory, enter ls -l to list all the files in that directory. Now run the VBoxLinuxAdditions.run file using the command:

./VBoxLinuxAdditions.run

Wait until all the modules successfully build and install.

https://imghostr.com/29eab4_ctc

Next, change back to the root directory and unmount the guest additions cdrom. After that, reboot the system by entering reboot in the terminal. Once the system has rebooted, you should be able to enter full screen mode.

2: VirtualBox Network Settings.

By default, the Kali Linux VirtualBox image uses the Network Translation Address (NAT) connection type. This connection type allows the guest system to connect to the network using the host system. However, the other devices on the network including the host system will not connect to the Kali Machine. For penetration testing, all machines in a network should be able to communicate with each other.

To change the system's network settings, follow the steps below.

Start by turning off the guest system. Then Open the VirtualBox Manager and select the Kali Linux virtual machine. On the right hand, select the network icon. On the dropdown box, select adapter 1 and change the Attached from NAT to Bridged Adapter. Then select the network interface to use for connections.

https://imghostr.com/240506_tk6

3: Activating Wireless Network

Running Kali Linux as a virtual machine does not allow you to use the embedded wireless device. However, you can use an external wireless network device.

To use the external network adapter in Kali Linux, insert your network adapter and navigate to Devices, USB Devices, and select the name of your network adapter. Once the network adapter connects, you can launch the terminal and type ipconfig to see the name and adapter information as shown below.

https://imghostr.com/23f257_5v1

How to Run Kali Linux in Live Mode

The other method of using Kali Linux is by running it live. To run Kali Linux live mode, burn the Kali ISO into a DVD or USB drive and boot with it. Select run Live.

One of the advantages of running Kali Linux in live mode is speed. Live mode is easy to setup and use. However, if you are running live mode, all the files and configurations settings made will be lost once you reboot your device. This method is very useful if you want to try out Kali Linux or perform a quick task. If you are going to use Kali Linux extensively, we recommend installing it on the hard disk.

Installing Vulnerable Machine

In this section, we are going to setup a virtual vulnerable server for testing purposes. This server will allow us to perform attacks discussed in later sections of the book.

A virtual vulnerable server is a great way to perform attacks on unauthorized servers on the internet. This virtual machine also gives us full control as we can increase and decrease its security according to the type of tests we want to perform. It also allows room for errors as we can just re-install the system in case of errors.

The vulnerable server we are going to use is Metasploitable 2. You can download Metasploitable file from the resource page below:

https://sourceforge.net/projects/metasploitable/

Once you download the file, extract it, and set it up using the steps outlined in the Kali Linux VirtualBox section.

https://imghostr.com/237683_mwl

Once configured, launch it and log in using username *msfadmin* and password *msfadmin*.

https://imghostr.com/b23436_fd1

NOTE: There are other vulnerable servers available you can use such as Metasploitable 3, Hera Lab, Pwn OS, Vuln Hub and others.

We are using Metasploitable 2 because it has an active development team Rapid 7 and tons of vulnerabilities in the application, network, and web and operating system layers. You can read more about Metasploitable 2 on the Rapid 7 page available from the resource below:

https://metasploit.help.rapid7.com/docs/metasploitable-2

Updating Kali Linux

Kali Linux has tons and tons of tools and applications that receive frequent updates. You should regularly update your system so that you get the latest features. Just like other Debian-based distributions, the package manager for Kali Linux is dpkg and therefore uses apt update command.

To update Kali Linux, open the terminal and enter *apt-get update*. This command refers to the sources.list file to get the repositories available. To ensure the correct repositories are available in this file, type the following command in the terminal, nano /etc/apt/sources.list.

The following repositories should be available:

https://imghostr.com/f770e3_hte

If these repositories are not included on your sources.list, you can manually add them and then type *apt-get update*.

The above command synchronizes the package's index files from the repositories included in the sources.list file. Make sure to synchronize all the repositories before installing or updating a software.

Once the packages have been synched, the next step is to update all the software and packages included in the system. You can use the following commands to update packages in Kali Linux

- *apt-get upgrade* – We use this command to upgrade only the packages included in version of Kali Linux we are using.

- *apt-get dist-upgrade* – We use this command to upgrade the entire Kali Linux distribution to the latest version.

Once you select the type of update you want to conduct, Kali Linux will perform upgrades automatically. The system may prompt for some input depending on the type of operation in performance.

NOTE: These types of upgrades may take a while depending on your machine's capabilities and your internet speed.

The following illustration shows an upgrade in progress.

https://imghostr.com/7ee52b_c55

Action Step

Use the materail you have learned in this chapter to setup your working environment in readiness for the practical Kali Linux lessons in later parts of this ultimate Kali Linux tutorial.

In the next part of the guide, we shall look at a basic introduction to working within the Linix kernel system, of which Kali Linux is, and its network interface so that as you move forth into using Kali Linux for pen-testing, hacking, and for your cyber security needs, you have a firm idea of the basic elements we shall be dealing with in later parts of this Kali Linux tutorial:

Part 2: A Basic Introduction to Networking and Working with the Linux Kernel System

In this section, we are going to cover the basics of working with the Linux kernel system and the working environment you set up above. This section is important for beginners because the Debian-based system upon which Kali works on is different from the common Windows or OSx interface.

Let us start by understanding the Linux filesystem and its organizational structure. A filesystem is the methods or data structure that an operating system uses to organize files and data to control how it is retrieved or stored.

Linux Directory Structure

All Linux based distro have similar structure even though some may have additional files and directories. In this section, we are going to use the Kali Linux directory structure.

https://imghostr.com/9edaa2_f6n

/- Root Directory

This is root or the starting point. Every directory in the Linux FS starts from this folder. In most Linux distributions, only the root user has access to this directory.

/bin/ - Binary Files

This is the directory containing basic programs

/boot/ - Kernel Boot Files

These are the Linux kernel and files required for the early boot process.

/dev/ - Device Files

This directory contains device files.

/etc/ - Configuration Files

This is the directory containing Linux configuration files.

/home/ - User Directory

This is the user's directory; it contains users' personal files such as videos, images, documents etc.

/lib/ - System Libraries

Directory containing basic libraries files.

/media/* - Mount Point

This is the mount point for removable devices such as USB drives, CD and DVD drives etc.

/mnt/ - Temporary Mount Point

This is the temporary mount point. It is available mostly in live Linux live mode.

/opt/ - Third Party Applications

Contains applications provided by third parties.

/root/ - Admin personal directory

This contains administrator's personal files such as media and documents.

/run – Volatile Data

Contains volatile runtime erased upon reboot.

/sbin/ - System Binaries

Contains system programs

/srv/ - Server Files

Contains data used by servers hosted on the system

/tmp/ - Temporary data

Contains temporary data erased at boot.

/usr/ - Applications

This is the directory containing application files. It has various other sub-directories.

/var/ - Variable data

Contains variable data required by daemons. These files include; log files, spools, queues, and caches.

/proc/

This is a special directory containing runtime system information e.g. system memory, devices mounted, hardware config etc. We can refer to it as a control center for the kernel during runtime.

/sys

Sys filesystem is a pseudo file system that exports device information and drivers through virtual files.

The Linux Command-Line

The power of the Linux operating system lies within an application called a terminal. It may have many names like terminator, terminal, iterm, xterm, and more but it is still a terminal.

Let us see what makes this tool such an important tool in the Linux/Debian-based operating system.

The Terminal

In Linux, the term terminal refers to an interactive, text-based interface. It allows the user to enter, execute, and view the results of the commands. In desktop-oriented distros, we can launch the terminal from a graphical desktop session while in server-oriented distros, it is the only thing you see when the computer boots up; server-oriented distros such as the Fedora server do not provide a graphical session.

Launching Terminal

By default, Kali Linux provides the gnome graphical desktop environment. If you downloaded the right Kali Linux and if your system is working properly, you should have a graphical desktop environment.

On the left hand of your desktop screen, you will notice a dock that contains a list of your favorite application. Select the black icon with a text inside; it's usually the second from top. You can also move your mouse pointer to your top left of the screen to activate the activity page.

Type terminal to launch it:

https://imghostr.com/58ccae_rd2

Command line Basics – Managing Files

As a Linux user, managing files via the terminal is very essential.

Rarely will you encounter cases where you switch from terminal to GUI control. In this section, we are going to see how to manage files from the terminal similar to how you would use a GUI (graphical user interface) file manager.

Launch the terminal and let us get started.

- ❖ pwd. This command stands for *print working directory;* it shows your current working directory in the file system.

- ❖ **cd** – We use this command to navigate to the targeted directory. The complete syntax is **cd <name of target directory>**. If the directory is outside the current working directory, the path is required. If you do not specify the target directory, it takes you to your home directory. To go back to the previous directory since your last **cd** call, type **cd ..** or cd –

- ❖ **ls** – We use the command **ls** to list all the files and directories within the current working directory. The syntax for **ls** command is **ls <option> [file|dir]**. The table below shows some of the options for ls command.

option	description
`ls -a`	list all files including hidden file starting with '.'
`ls --color`	colored list [=always/never/auto]
`ls -d`	list directories - with ' */'
`ls -F`	add one char of */=>@| to enteries
`ls -i`	list file's inode index number
`ls -l`	list with long format - show permissions
`ls -la`	list long format including hidden files

- ❖ **mkdir** – We use this command to create new directories. To create a new directory in your current workspace, type **mkdir <dir name>** For example, **mkdir newFolder.** For cases where you want to create a directory outside your current directory, you must specify a path. For example, **mkdir /user1/Desktop/Report/lastWeek** where in this case, lastWeek is the name of the directory you want created.

- ❖ **rmdir** – We use the **rmdir** command to remove directories. Similar to the **mkdir** command, this command takes options that determine how the system deletes files.

- **mv** and **cp** – We use the **mv** command to move and rename directories and files. We use the **cp** command to copy files. The syntax for copying files is **cp <source file> <target-file>**. Here is an example, **mv helloDir > /usr/share/**.

Other Useful Commands

Apart from the file management commands we have discussed above, Linux supports a wide array of commands.

In this section, we are going to look at some of the must know.

COMMAND	Description
help	This lists all the commands in the terminal.
tar	We use this to extract, create, and view tar files.
find	We use this command to search files matching certain criteria in a specified location.
gzip	We use this to command extract and create gzip archives
unzip	We use the inzip command to extract or unzip a zip archive.
whatis	This command shows the use of a certain command; it gives a short description of a command
Man	This command shows the manual of a specified command.

exit	The exit command ends the current bash session. We can have multiple bash sessions in one terminal.
sudo	We use this command to run a specified command as the superuser or root user.
su	In the terminal, we use this command to switch current user to a specified one.
top	This command acts as a task manager in terminal and shows top processes.
shutdown	This command shuts down the computer after a specified time.

NOTE: Linux Commands are case Sensitive. If you type them incorrectly, they will not work. If you want to master Linux, Master the shell.

Networking Basics

To understand networking in in the Debian-based system, you need to understand the following concepts:

Network Hosts

A network host, simply called host, refers to any devices connected to a specific network. It could be a printer, personal computer, mobile device or even a web server.

Network Protocols & Services

The following are some of the most common protocols:

1: Internet Protocol

Internet protocol is the main set of rules that are very essential in computer communication. We normally refer to this as the IP address. It is unique set of characters identifying each device connected to a network. There are two main types of Internet Protocol:

Private IP address

A private IP address is a type of IP that identifies devices connected on a private network such as LAN. They range from 1 to 255. An example of a typical private IP would be something like ***192.168.0.2*** or ***192.168.0.220***

To find the Private IP of your device on your local network, open up the command prompt and type ***ipconfig*** and ***ifconfig*** on windows and Linux respectively.

https://imghostr.com/603f39_64f

Public Internet Protocol

A public IP address is the other type of IP used to identify devices connected on the internet. Public IP is globally unique. To find your public IP, open your browser, and using Google Search, type and search for "what is my public IP."

2: HTTP & HTTPS

HTTP and HTTPS are another common type of network protocol that refer to Hyper Text Transfer Protocol (HTTP) and Hyper Text Transfer Protocol Secure (HTTPS) respectively. This type of protocol allows communication between web servers and typical web browser. It is the second most used protocol.

HTTP and HTTPS only differ on security. The information transferred on the HTTP protocol is unencrypted and easily accessible through sniffing methods and such. HTTPS uses secure socket layer (SSL) to encrypt the information.

https://imghostr.com/8e017f_aou

3: Secure Shell – SSH

Secure shell is becoming a very common protocol among network administrators and IT professionals. It allows the connection of remote hosts for communication.

Suppose you have a dedicated server and you want to install programs on it. With the use of SSH client and the SSH server on the remote hosts, you can be able to gain access to it and configure it as if you were actually there to configure it.

SSH is similar to the Microsoft Telnet protocol except that it is more secure and more powerful compared to Telnet. This protocol is most common among Linux users. Connecting to SSH requires a *username* and *password*.

4: File Transfer Protocol

File Transfer Protocol, also called FTP, is another common type of network protocol. It allows upload and download of multiple files to and from a remote host. Developers and IT professionals normally use it to upload and download files on the server.

Connecting to the server requires an FTP client and a *username* and *password*. The most common FTP client is FileZilla; you can download FileZilla from the resource page below:

https://filezilla-project.org/

Network Ports

Network ports and network protocols are very similar. We normally say that a computer running a protocol such as HTTP or FTP is running the HTTP service. These services or protocols associate with a specific number called a port. A single port can support only one service on a single machine. If another device wants to access a specific service on a certain machine, it has to connect to the specific port where the intended service is running

A simple explanation can be that ports are doors where each door allows access to a specific service within the computer. Network ports are very important in hacking since exploiting a certain service on a computer can allow further access.

The screenshot below shows the list services running on a windows 10 machine.

NOTE: You can configure these services to use a specific port; they are therefore un-fixed.

https://imghostr.com/a1d25d_9xn

Defining which port a service will be running on within a specific computer is very difficult because a computer has over 65, 535 TCP ports and 65, 535 UDP ports all of which are configurable to run any service.

Action step

This part of the guide has covered everything you need to know about working within the Kali Linux interface and how the networking system works within the system. Go over these elements several times so that you internalize them and can therefore follow along as we delve deeper into learning how to use Kali Linux.

Part 3: Information Gathering Tools

The first set of tools Kali Linux offers are information-gathering tools. Information gathering, also called reconnaissance or recon, is the very first step in penetration testing or launching an attack against a certain target.

In the information-gathering toolset offered by Kali Linux, we are going to look at the most important ones that include the following:

- ❖ Nmap and Zenmap
- ❖ Netdiscover
- ❖ DNS Tools
- ❖ H3Ping
- ❖ Load Balancing Detectors

NMAP & Zenmap

Nmap is the often referred to as King of scanners. It is the most advanced tool for performing network scanning. Zenmap is the Graphical version of Nmap as it is a command line tool. Nmap offers advanced features in network discovery and auditing.

In cyber security, we normally use Nmap to detect the devices connected on the network, determine services running on them, versions of the specified services, and even the operating system running on the target system. It uses raw packets to perform the specified tasks.

How to use Nmap in Kali Linux

Here is how to use Nmap in Kali Linux.

Fire up your Kali Linux and go to applications – information gathering – Zenmap. We are going to use the GUI version of Nmap as it is easy to use and comfortable for people with not much command line experience.

Discovering Hosts

The first step in the information-gathering phase is to determine the connected clients. A network can support over more than 200 clients. Scanning the information on all of these devices can take a while. Nmap offers a very efficient way to perform host discovery in order to find your target device on the target network.

Let us start up Kali Linux and Metasploitable to perform host discovery.

https://imghostr.com/6fdf38_u95

Once you have launched Zenmap, enter your network gateway/24, which scans the entire network subnet. The illustration below show nmap intense scan of the entire network:

https://imghostr.com/4291d2_wmi

Do not worry about most of the details in the above illustration. We can see that once the network scanning process completes, the process discovered 7 hosts, two of which were discovered as Linux based system. The details we are looking for is the Metasploitable machine. We discovered the IP address of Metasploitable and the services running on it.

Stealth Scans

Stealth scan is also known as SYN scan. An SYN scan works by sending a SYN packet to the target host. According to the general rule of networking, once a system sends a SYN packet, the targeted device will return an ACK (acknowledgment) response. If this does not happen, Nmap uses this technique to determine the open and closed ports in a target system.

To perform more detailed scan on a target, instead of scanning the entire network, enter the IP address of the target system in Zenmap.

The following result is what the system returns for Metasploitable.

https://imghostr.com/3b6a23_bml

Netdiscover

Netdiscover is another tool bundled into Kali Linux. This lightweight tool allows us to scan the network for connected devices quickly. It is very useful when doing a quick scan and discovering the IP and MAC addresses of connected devices on the network. To perform a quick scan, open the terminal and enter *netdiscover* command. You can also use the help flag in Linux to find more command options.

https://imghostr.com/a11673_guk

DNS Tools

DNS tools help in zone transfers and IP resolving for domain names. In this section, we are going to look at some of the DNS tools Kali Linux has.

Dnsenum

Dnsenum is a tool used to get records of a stored domain such as MX and A records. It is a Perl script and requires that you install Perl in Kali Linux. To use dnsenum, type

<dnsenum domainname.com> on the terminal; the following illustration shows a dnsenum for facebook.com:

https://imghostr.com/9cd401_p2n

DNSMAP

We normally use this tool to perform a DNS mapping of a specific domain name. We can use it to extract information such as phone numbers, contact information, and other subdomains within it.

HPing3

H3ping is a more advanced ping tool used by a wide range of hackers. It allows the attacker to perform ping by bypassing firewall filters. It uses RAW-IP, TCP, UDP and ICMP protocols for ping requests. H3ping supports file share between channels and traceroute mode.

You can use the help flag in Linux to find more information on the command parameters. The basic command is <hping3 ip-address>

Action step

The tools we have discussed here are just a serving of the many information-gathering tools within Kali Linux. As you use these primary tools, your ability to use the various other information-gathering tools within Kali Linux will automatically improve.

In the next part of the guide, we shall look at how to use the primary vulnerability assessment tools within Kali Linux

Part 4: Vulnerability Assessment

In hacking, we use Vulnerability assessment tools to recognize, categorize, and characterize security vulnerabilities within a system. These tools automate vulnerability scanning and security audit.

In this section, we are going to look at some of the top vulnerability assessment tools in Kali.

Web Application Scanners

When developers do not adhere to secure coding principles and techniques, a web application can contain vulnerabilities that allow an attacker to penetrate them. Here are some of the tools used to scan web applications for vulnerabilities.

Nikto

Nikto is web vulnerability scanner bundled into Kali Linux. Nikto is very powerful and supports a wide range of capabilities. To launch nikto, navigate to Applications – vulnerability scanners – Nikto. You can use the help flag to find more information on command parameters. Nikto can detect outdated server implementation and misconfigurations.

OWASP ZAP

OWASP is an open-source, web-application proxy tool used to perform web application vulnerability scanning. You can launch it in Kali Linux by entering the command *owasp-zap*.

OWASP offers a graphical window to work with. Once the window has launched, you can enter the URL of a web application and click attack. OWASP will report all misconfigurations of the application in the alerts tab.

https://imghostr.com/394397_lz1

W3af

W3af is a python web scanner used to identify and exploit all web application vulnerabilities. This web-application audit framework offers both graphical and command line usage. We normally refer to it as 'Metasploit for web application' and we can use it to perform attacks such as cross-site scripting, remote file inclusion, SQL injections etc.

http://w3af.org/

Network Vulnerability Scanners

Hackers and cyber security experts normally use network tools to scan routers, databases, protocols, and switches. Kali

Linux offers a ton of tools for network scanning but we will only look at the most common one.

Cisco Tools

CISCO devices are usually prone to a number of vulnerabilities that are accessible using a number of tools. Here are some of the tools we are going to cover:

Cisco Auditing Tool

This is a Perl script tool that helps a penetration tester perform security audits and identify vulnerabilities such as SNMP strings, cisco bugs and default credentials for cisco devices. The syntax for launching it is *CAT -h IP -w lists/community -a lists/passwords -i*

Cisco Torch

We use Cisco torch to perform mass scanning, fingerprinting, and exploitation of vulnerabilities within cisco routers. You can launch it by typing *cisco-torch -parameters IP-address*. Once the command has executed, the tool will report if it has found anything to exploit. You can use the help flag to find more on the parameters required.

Cisco Global Exploiter

CGE is another cisco exploitation tool. We normally use this advanced and simple tool to perform attacks on cisco devices. You can launch it by typing *cge.pl IP-address*. Wrong use of this tool such as performing Router Denial of Service can cause devices to crash.

Action step

Before moving on the next part of the book, orient yourself well with these, and the other vulnerability assessment tools within Kali Linux:

Part 5: Wireless Attacking Tools

In this section, we are going to look at some of the wireless hacking tools offered by Kali Linux. Our primary focus shall be Wi-Fi hacking tools. To use these tools, you need a wireless card with monitor mode support.

Kismet

Kismet is a 802.11 layer wireless network analyzer, detector, sniffer, and IDS tool. Kismet supports a wide range of wireless cards provided they support monitor mode. We normally use the tool to sniff 802.11a/g/b/n traffic and intrusion detection.

To use it, you need to turn your wireless network card into monitor mode. To discover the name of your wireless card, type *ifconfig* command in the terminal; in most cases, it will be **wlan0.** Once you get the name of your wireless card, enter the command **airmon-ng start wlan0**

- ❖ Launch the terminal and enter kismet command
- ❖ Click OK on the window that appears
- ❖ Agree to start the Kismet server upon request otherwise it will not work
- ❖ Continue with default startup settings

- Next add your wireless card
- Click Ok to start sniffing for wireless networks

Wifite

Wifite is a network-cracking tool used on WEP, WPA and WPS networks. To use it, you need to set your network card to monitor mode using the command shown above.

Launch wifite by navigating to applications – wireless attacks – wifite. To show the networks available, use the command wifite -showb. Once you have found the network to attack, end the scan by hitting CTRL + C. select the target by entering its corresponding number. The tool will start cracking the network and display the key once it finds it.

Fern Wifi Cracker

Fern Wifi cracker is graphical wireless cracking tool. It is very easy to use and is therefore very friendly for beginners. Launch it by navigating to applications – wireless attacks – fern.

- First select the wireless card you want to use – ensure the card is in monitor mode
- Next, click scan for access points to scan for available networks.

- ❖ Once the scan has completed, it will display all the available networks.

- ❖ Once you select the network to attacks, click Browse to choose a wordlist to use for the brute-force attack.

- ❖ Once you select the wordlist, click 'attack" to start the brute-force

- ❖ If the tool finds the key, it will display it below the program

Action step

The more you experiment with these tools, the better you will get at using them. Before you move on to the next part of the book, practice using the various wireless attacking tools within Kali Linux.

Part 6: Web Application Attacks

We previously looked at some web-application vulnerability scanners such as ZAP and Nikto. In this section, we are going to look at other web attacks such as databases, and CMS such as WordPress and Joomla.

Database Tools

Tools in this category are especially effective at performing database vulnerability checks and exploitation. The most popular tools in this category are:

SQLMap

SQLMap is an open-source database assessment tool that automates the detection and exploitation of SQL injections allowing for the compromise of database servers. SQLMap has a detection engine that allows it to offer features such as data fetching, database fingerprinting, database filesystem access, and database command execution.

This tool supports databases such as MySQL, PostgreSQL, Oracle, SQLite, Microsoft SQL server, IBM and Max. SQLMap is a python-written program, and is one of the most powerful and popular SQL injection automation tool. The illustration below shows a SQLMap attack on Mutilidae vulnerable web application in Metasploitable 2.

Check out more on www.sqlmap.org

https://imghostr.com/4f97e0_wzm

https://imghostr.com/793589_5mf

SQLNINJA

SQL ninja is database exploitation tool used to exploit web applications that run Microsoft SQL server as its backend database. This tool creates a shell on the remote host once it discovers a SQL injection. You can check out more information on SQLNINJA by reading the manual documentation or by visiting the resource page below:

http://sqlninja.sourceforge.net/

Vega

Vega is an open source testing and scanning platform written in Java for web applications. Vega helps in identifying and validating SQL injections, XSS, and other vulnerabilities in a web application. It can also be included in Web via javascript apis. More information is available on the following resource page:

https://subgraph.com/vega/index.en.html

NOTE: Kali Linux has discontinued Vega from their latest releases.

CMS Tools

CMS, or Content Management System, is a system that manages the development of digital information. Some of the most popular CMS include WordPress, Joomla, and Drupal. In this section, we are going to look at some of the tools Kali Linux has for finding and exploiting vulnerabilities in CMS.

WordPress Scanner

WordPress Scanner or WPScan is a WordPress vulnerability scanner used to scan WordPress installations for vulnerabilities in themes, plugins, and databases. You can launch it by typing **wpscan** command in the terminal.

https://imghostr.com/49f1e2_txe

You can provide other parameters to include in the scan. Once the scan has completed, wpscan will list all possible vulnerabilities as shown below:

https://imghostr.com/39067a_g4v

JoomScan

Because of its flexibility, Joomla is the second most popular CMS. JoomScan is a tool for scanning Joomla-based websites for possible security flaws. Developers and web gurus normally use this tool to find errors in the web applications.

Here is how to use the tool:

- ❖ To launch JoomScan, open the terminal and enter joomscan -params

- ❖ You can use the help flag to find more information on required commands.

- ❖ To scan a certain website use **joomscan -u target-URL**

Action step

Orient yourself with using the various web-application attack tools we have discussed in this part of the guide:

Part 7: Exploitation Tools

Exploitation is the art of using vulnerabilities in a system to gain unauthorized access. Once we discover a vulnerability using some of the tools discussed above, we use exploitation tools to take advantage of the vulnerability and gain access to the system.

In this section, we will discuss the most common exploitation tools bundles in Kali Linux.

Metasploit

Metasploit is the most popular exploitation framework equipped with thousands of scripts. This penetration framework helps in finding, validating, and exploiting vulnerabilities in computer systems. Developed by Rapid 7, this tool offers a command line interface for operations.

Metasploit contains scripts written in different languages for platform hacking. It is also cross platform and can run on Linux, Windows, and OSX. You can find more information from the following resource page:

https://metasploit.help.rapid7.com/docs/getting-started

NOTE: Metasploit is a very complex and advanced framework.

To launch Metasploit in Kali, open applications – exploitation – Metasploit framework

https://imghostr.com/17e91d_wbv

When you launch Metasploit for the first time, the interface can be very intimidating. Let us discuss the basics of working with Metasploit that you need to know to get started.

Metasploit Fundamentals

MSFconsole is the most popular interface for interacting with the Metasploit framework. It provides a terminal control of the entire framework. Once you have launched MSFconsole as shown above, a banner, and information about what Metasploit contains will appear.

To get started, type the command **help** to list all the commands supported by the Metasploit console. The Metasploit console has load of commands.

Here are the most common and essential ones.

- ❖ Back: returns you back to your previous context
- ❖ Banner: randomly selects and displays a banner
- ❖ Show options: displays the options and requirements of a specified module

- Search: search for modules based on the name, description, and platform
- Route: routes network traffic through a Metasploit session
- Use: selects a module by its name
- Kill: used to kill a running job
- Load: used to load a plugin into the framework
- Connect: used to establish communication with the host
- Sessions: displays detailed information about running Meterpreter sessions.
- Set: Sets a specified module variable to a value
- Unset: unsets a specified variable from its current value.
- Info: displays detailed information about specified modules.
- Help: shows help on Metasploit commands and usage.
- Irb: drops into a ruby shell

These are some of the available Metasploit commands. Keep in mind that Metasploit supports more than 200 commands.

We are going to look at an example of how to use Metasploit to exploit Metasploitable 2 VSFTPD vulnerability.

https://imghostr.com/3bf9fe_qiv

The first step is to search for the vulnerability exploit. Load the module by using the use command, followed by the full exploit name.

The next is to show information about the exploit. In this case, it is a VSFTPD backdoor command execution.

https://imghostr.com/6e2dd4_oq5

Next, set the required arguments displayed by the show info command. The RHOSTS in this case is the IP address of the target machine (Metasploitable 2). The next is to set a payload. A payload is set of code or command to be executed once the vulnerability has been exploited. Once you set the payload, check the info to make sure no required parameters are missing. This in place, the next step is to exploit the target using the exploit command.

https://imghostr.com/051d42_ny6

Once the target has been exploited, depending on the payload used, it will give you a terminal access or a Meterpreter session. In this case, a terminal access as root was given.

NOTE: As noted earlier, Metasploit is very advanced and proficiently working with it requires advanced knowledge that can only come from consistent practice.

Armitage

Armitage is graphical version of Metasploit. This Java-written program provides a very handy and easy-to-use interface for working with Metasploit. It provides GUI mapping of targets and visualizes recommended exploits, payloads, and post-exploitation commands.

Kali Linux Armitage Setup

To setup Armitage without errors, perform the following commands.

- Service PostgreSQL start – used to start the database

- Service Metasploit start – used to start the Metasploit service and initialize the database

- Service Metasploit stop – used to stop the Metasploit service

- update-java-alternatives --jre -s java-1.7.0-openjdk-amd64 – updates the java version

Once you execute the above commands, launch Armitage and click connect.

https://imghostr.com/3b0cb7_zpd

To start using Armitage, you can add targets by clicking on the hosts tab.

https://imghostr.com/f2275a_5ic

Once you have added the targets, you can start perform operations such as scanning, finding vulnerabilities, and setting payloads.

https://imghostr.com/8718c7_6yw

You can find more information on how to use Armitage on the following resource page:

http://www.fastandeasyhacking.com/

Browser Exploitation Framework (BeEF)

BeEF is another penetration-testing tool used by security professionals. BeEF exploits browser-based vulnerabilities to perform client-side attacks. It is open source and comes bundled in Kali Linux. You can find more information about this tool from the resource page below:

https://beefproject.com/

To launch BeEF, open the terminal and navigate to /usr/share/beef-xss

https://imghostr.com/688eff_fdw

Launch BeEF by executing the .beef file in the directory. Note two URL provides by BeEF: UI URL and Hook URL.

https://imghostr.com/1bee9e_mw8

Open the UI URL in the browser. Enter the username and password to login into BeEF control panel. In the latest versions of BeEF, you have to change the default password in /usr/share/beef-xss/config.yaml for the server to work.

https://imghostr.com/3f34ef_p4a

The Hook URL is a JavaScript file hosted by the BeEF server and runs on client browsers hooking it to the BeEF server. Once the hooked JavaScript file executes on the target machine, you can control the machine's browser.

Write a simple HTML file and include the hook URL in the file. You can now send the file to the target and once the target opens it, it will hook the target's browser and will appear in the Online browser's tab in the BeEF control panel.

https://imghostr.com/60528a_9mw

Exploit Database

Exploit database or exploitdb is an archive of searchable exploits and vulnerabilities developed and maintained by offensive security. It contains scripts and detailed information about discovered exploit. You can launch it by typing searchsploit in the terminal. You can also visit the web page below for easier navigation

http://exploit-db.com

Action step

As has been the case with all the other Kali Linux pen-testing tools we have discussed in the guide, mastery of suing these exploitation tools will only come from practice.

Practice using the various exploitation tools described here:

Part 8: Forensics Tools

Forensics is becoming a very important aspect of cyber security. It is the art of using scientific methods to investigate a crime and perform recovery measures for evidence. Kali Linux has tools to help you achieve this easily.

Let us look at some of the Kali Linux tools used in this area.

Autopsy

Autopsy in an open-source forensic tool used to investigate disk images. It provides a nice browser user interface for easy control and management of case files. Usage of this tool is common among military or law enforcement personnel who use it to perform digital forensic operations.

Autopsy contains a wide range of functionalities such as keyword search, time analysis, hash filtering, data craving and compromise indicators, multimedia extraction, and web composition extraction such as cookies, history, and bookmarks. It allows one to sort files according to Case Name, Description, and Investigators Name.

Autopsy supports disk images in certain types of formats such as E01 and Raw images. It then produces reports in HTML and XSL formats. It is a robust tool used in the management and analysis of digital forensics. You can learn more about autopsy from the resource page below:

https://www.sleuthkit.org/autopsy/

To launch autopsy, navigate to Applications – Forensics – Autopsy. Once you launch it, you can access its web-based interface by opening:

https://localhost:9999/autopsy

https://imghostr.com/5d6447_en2

Binwalk

Binwalk is another forensics tool written in python that makes it possible to analyze binary images. It allows penetration testers to find embedded files and executable files. It provides very powerful features to those who know how to use it expertly.

You can use the tool to extract sensitive firmware information that you can then use to uncover a possible hack. It utilizes the libmagic library, which makes it efficient at working with UNIX signature files.

You can find more information from this resource:

https://github.com/ReFirmLabs/binwalk

https://imghostr.com/20d606_wvc

Rkhunter & Chkrootkit

Rkhunter, or rootkit hunter, is a program used to find rootkits in installed systems. It allows detection of compromised systems and removal of malicious files in a computers system. Chkrootkit is similar to rkhunter and can be able to detect system binaries for malicious modifications, deletion of log files, string replacements, and temp file deletions.

Both programs provide very simple interfaces.

https://imghostr.com/df400f_gbn

Volatility

Volatility is the most popular python framework for extracting digital data from RAM samples. Volatility supports memory dumps in several formats such as crash dumps, hibernation files, and VM snapshots (a copy of virtual machine disk file). Volatility is compatible with 64 and 32-bit variations of windows and Linux distros such as Android.

This tool comes in handy when investigating encrypted data on a hard disk. Since this kind of data is stored in RAM, when available, this tool saves a lot of time and makes tasks easier.

Find more information on this tool from this resource page:

https://www.volatilityfoundation.org/

https://imghostr.com/5588a0_urc

Hashdeep

Hashdeep is hash-auditing tool that deals with SHA-256 and MD5 hash formats. It performs recursive hash calculations using various algorithms.

https://imghostr.com/ac31e5_owh

DDRescue

DDRescue is a hard disk data extraction tool used to copy data from a hard disk block in case of read error. DDRescue performs automatically; it reads data from good blocks of a hard disk and recovers it. This tool is quite advanced; do not use it unless you know what you are doing. It may lead to data loss or disk damage.

Action step

Practice using the six tools we have discussed here; they are the one commonly used in forensic auditing and their mastery will enrich your forensic testing knowledge and expertise:

Part 9: Sniffing and Spoofing Tools

In this section, we are going to discuss sniffing and spoofing techniques using tools provided by Kali Linux. In some instance, we will deviate a little from the tool's topic and into networking.

NOTE: To understand this section, you need a proper understanding of networking basics such as protocols, DNS, and OSI model. If you need to, cycle back to the part 2 of the book to relearn the most important networking basics:

ARP Protocol

ARP protocol or Address Resolution Protocol is a communication protocol used for discovering link layer address of a given layer address. We normally use it to map network addresses such as Internet Protocol v4 to its respective physical address. You can read more about ARP protocol from the following resource page:

https://en.wikipedia.org/wiki/Address_Resolution_Protocol

How ARP Works

The ARP protocol is similar to a DNS. If you are unfamiliar with how DNSs works, a DNS matches a domain name to its respective IP address. The same applies to the ARP protocol that maps an IP to an unknown mac address.

Take the example of a device with an IP address 192.168.0.20. This device wants to communicate with a device on the same network with IP address 192.168.0.21 but does not know its physical address (MAC address).

The .20 IP address device will broadcast an ARP request across all devices on the network asking who owns the IP address 192.168.0.21. When the device with this matching IP address receives this request, the owner of the requested IP will reply with an ARP response that contains the physical address of the device. This information is cached in the ARP cache containing the IP address and its respective Mac address.

Next, we are going to perform sniffing and spoofing using the following tools

- ❖ Ettercap
- ❖ Wireshark
- ❖ SSL strip

Ettercap

Ettercap is network sniffer tool used to perform spoofing and man-in-the-middle attacks. In this section, we are going to see how to perform ARP poisoning using Ettercap.

Launch Ettercap from the Kali Linux application menu. Once open, navigate to the top bar and select, sniff. Select unified sniffing and then continue to select your network interface such as eth0 for Ethernet and wlan0 for wireless network.

https://imghostr.com/f5f0e8_iyc

Next, select the hosts tab and click scan for hosts. Once the scan is complete, you will see the list of connected devices on your network.

https://imghostr.com/35dc71_2pz

Next, we need to specify target1 and target2. Target1 is the target machine we want to attack; target2 is the IP address of the router.

With the targets specified, open the MITM (man in the middle) and click ARP poisoning. Next, select sniff remote connections and click OK. Next, in the start menu, select start sniffing to start the process.

https://imghostr.com/583175_12z

This process involves Ettercap sending packets to the network with requests to update the ARP cache replacing the routers MAC address with the MAC address of your computer. This allows us to view all the unencrypted information passing on the network. This included passwords and other sensitive information. You can use Ettercap to perform other operations such as DNS spoofing.

Wireshark

Wireshark is a popular network tools bundled with Kali Linux. It is a packet sniffer and analyzer in the frame level. For better understanding of how Wireshark works, we are going to cover the OSI model in basics.

The OSI model

The Open System Communication also called OSI model is a model that regulates the communication between devices on the network. The transfer of data between devices within the same networks occurs with seven layers of this model with each layer having a specific role transmitting data to or from the next layer.

A deeper explanation of the OSI model is below:

When we are sending data from a specific device, the process of transfer occurs in the following order:

Application Layer – Presentation Layer – Session Layer – Transport Layer – Network Layer – Data Link Layer – Physical Layer

The data process occurs in reverse, i.e. it starts from Physical Layer to Application Layer, when a device on the network is receiving data.

Let us discuss the functions of each layer in the OSI model.

- ❖ Physical Layer – This layer sends and receives raw bits of data over a physical medium such as Ethernet cables, Wireless cards, etc.

- ❖ Data Link Layer – We normally use this layer to broadcast data frames between physically connected nodes —it includes the MAC address.

- ❖ Network Layer – We use this layer to split data frames in packets by adding routing information

- ❖ Transport Layer – This layer is responsible for connecting two nodes for data transfer.

- ❖ Session Layer – This is the layer used to handle communication sessions between transmissions.

- ❖ Presentation Layer – This layer is responsible for data security such as encryption and decryption of data for application layer.

- **Application Layer** – provides network services to users by use of software applications. They include HTTP, FTP, SMTP, HTTPS, DNS etc.

For a deeper understanding of how the OSI model works, navigate to the resource page below more information

https://en.wikipedia.org/wiki/OSI_model

To learn more about Wireshark, see the page below:

https://wiki.wireshark.org/

Once you launch Wireshark, select the network interface, and then click on start sniffing. Once the tool captures network traffic, it will display the information as illustrate below:

https://imghostr.com/227a57_pfz

From the above image, we can see that Wireshark captured traffic in the second Layer containing mac address and IP address. You can find more information on how to read and analyze network packets here:

https://www.hackingarticles.in/network-packet-forensic-using-wireshark/

SSL Strip

SSLstrip is a MITM tool that converts HTTPS traffic to HTTP to allow the attacker to view information in plain text. It strips https://www.url to http://www.url

You can start it by typing SSLstrip in terminal.

https://imghostr.com/fb7ef2_6mv

Action step

Kali Linux has various other sniffing and spoofing tools but the ones we have discussed in this part of the guide are the primary ones that you should know how to use:

Part 10: Password Cracking

Passwords are a very common way of securing digital information and accounts. A password is a sort of 'secret code' or 'word' used for owner-verification to gain access to personal information. In this section, we are going to cover ways to use Kali Linux tools to crack passwords.

We can define Password cracking as the art of recovering passwords from previous stored locations and formats. For this section, we are going to cover the common set of tools for password cracking using Kali Linux. They include:

- ❖ John the Ripper
- ❖ Medusa
- ❖ Hydra
- ❖ Crunch
- ❖ Rainbow crack
- ❖ John

John the Ripper

John the Ripper is a popular, open source password-cracking tool from the Rapid 7 class family. John the Ripper supports brute forcing, dictionary and single crack attacks. It automatically detects password flaws that make the computer system vulnerable. You can learn more about john the Ripper from the resource page below:

http://www.openwall.com/john/

In this example, we are going to extract windows passwords stored in the SAM database. Follow the steps below.

- ❖ Download and install PWDump and extract it. Download
- ❖ Navigate to the extracted directory and enter the command shown below
- ❖ PwDump7.exe > d:\hashed.txt
- ❖ After the password hash completes the extraction process, open John the Ripper and enter the following command john --format=LM path/to/hashed_file.txt
- ❖ After the tool cracks the password, it will display on the tool corresponding to its username.

Medusa

Medusa is a network password-cracking tool that provides modular and parallel brute force. Medusa supports a wide range of protocols such as Telnet, SSH, PostgreSQL, MySQL etc. It requires a target IP address, username wordlist, and password wordlist. It offers a command line interface.

You can launch it by entering the command medusa in the terminal, which returns the list of possible commands.

How to Use Medusa to Crack Usernames and Passwords for an Ftp Protocol

Let us look at an example of how to use medusa to crack usernames and passwords for an ftp protocol.

medusa -h 192.168.0.25 -U ~/path/to/usernames.txt -P ~/path/to/passwords.txt -M ftp

https://imghostr.com/a617dc_an4

Once the tool finds the matched username and password, it appeas displayed as shown below:

https://imghostr.com/4e3421_dh7

Crunch Password Generator

Crunch is a simple password generator tool. We normally use it to generate random usernames and passwords and store them in a file. It accepts a min and max length followed by the characters to be included in the wordlist. Crunch supports the following features:

- ❖ Unicode characters
- ❖ It supports 'resume of password generating process'
- ❖ It supports numbers and special symbols
- ❖ It uses both combination and permutation ways
- ❖ It supports lower and uppercase characters

https://imghostr.com/9b5e97_wqf

Once you set the min and max value, you can then set where crunch will save the file once it generates the wordlist.

https://imghostr.com/84fe16_ar1

Rainbow Crack

Rainbow crack is password-cracking tool that uses rainbow tables. Rainbow tables are password files that you can download from online sources and store in the computer storage. To launch rainbow crack, open the terminal and enter the following command:

rcrack   ~/path/to/rainbow_table   -f ~/path/to/password_hash

Rainbow tables are available in different formats such as NTLMM hashes, MD5, LM hashes, SHA-1

You can download a rainbow table from the following resource:

http://project-rainbowcrack.com/table.htm

NOTE: Some rainbow tables range from 100GB to 10TB. The larger the rainbow table, the longer it takes to crack the password.

Johnny

Cyber security experts and hackers rarely use this tool. The tool is a graphical interface of John the Ripper, a tool discussed earlier.

https://imghostr.com/06599d_0np

Action step

The secret to mastering these tools —and any aspect of cyber security or hacking for that matter— is to practice using them as often as possible so that you can internalize how to use each tool effectively. The next part of the guide looks at the various post exploitation tools within Kali Linux

Part 11: Post Exploitation

Once you compromise and gain access to a computer system, the next task is usually to maintain access to this system. There are a million of ways to lose access such as updates, detection of your presence in the system, antiviruses, and such.

In this section, we are going to look at the various methods you can use to maintain access to a system you have successfully compromised.

Weevely

Weevely is tool used to generate stealth PHP backdoors providing shell controls into the system. To launch Weevely, open the terminal and type **weevely** to see the available commands.

How to use Weevely to exploit Metasploitable 2 DVWA web application

Here is an example of how to use Weevely to exploit Metasploitable 2 DVWA web application.

Step 1: Launch Metasploitable 2 and Kali Linux

Step 2: Open the terminal and type Weevely generate <password> ~/path/to/save_file.php

https://imghostr.com/283550_e5l

Step 3: Open your Kali Linux web browser and navigate to <ip_of_metasploitable>/dvwa and login with username: admin and password: password.

https://imghostr.com/68669b_0xh

Step 4: Navigate to upload tab and select the file generated by Weevely and click upload.

https://imghostr.com/9b2f0c_swn

Step 5: Once the file completes uploading, navigate to the backdoor.php file to execute it.

Step 6: The next step is to open Weevely and enter the command as Illustrated on the screengrab below:

https://imghostr.com/a82095_5r7

The syntax above is Weevely <url_of_file.php> <password>

Step 7: Once the file connects, you will get a backdoor access to the web server.

Powersploit

Powersploit is an exploitation tool that allows a hacker to install a PowerShell backdoor to the target machine, thus allowing you to connect to the target machine via PowerShell.

Powersploit has modules and scripts such as Antivirus Bypass, reverse engineering, and reconnaissance. You can find these modules by navigating to /usr/share/powersploit in Kali Linux.

You can learn more about powersploit from the following resource page:

https://github.com/PowerShellMafia/PowerSploit

Shellter

Shellter is a very popular shellcode insertion tool that encodes payloads to bypass some security software applications such as Antiviruses. It works by embedding 32-bit windows applications with the shellcode thus evading antiviruses. You can use shellcodes from Metasploit and other frameworks or create your own.

- ❖ Install Shellter using apt-get install Shellter
- ❖ Next install wine using apt-get install wine32
- ❖ Next, launch Shellter from the terminal
- ❖ Shellter provides an easy to use interface with interactive questions

Nishang

Nishang is another post exploitation tool similar to Powersploit offering PowerShell scripts to use alongside Metasploit.

You can find more information on Nishang from the resource page below:

https://github.com/samratashok/nishang

Action step

Practice using the various tools discussed here. Other parts of this guide shall have practical practices that will test your ability to use some of the Kali Linux tools whose usage of we are illustrating in this guide. The more you practice, the easier and more intuitive using Kali Linux becomes, and the better a hacker/cyber security analyst you become.

In the next part of the guide, we move on to social engineering:

Part 12: Social Engineering

Social Engineering is the art of psychologically manipulating a target client into performing tasks that lead to you gaining access to otherwise private information. It is a very powerful and popular technique among professional hackers. In this section, we are going to cover some of the Kali Linux tools that allow us to perform these kinds of attacks.

Social Engineering Toolkit (SET)

Social Engineering Toolkit, or SET for short, is the most popular social engineering tool. By design, the tool's main task is to perform attacks on human beings through psychological manipulation. SET offers a wide range of features such as:

- ❖ Infectious Media
- ❖ SMS spoofing
- ❖ Web Attack Vectors
- ❖ Human Interface Attack Vectors

To launch SET in Kali Linux, open the terminal and enter the command **setoolkit**. This provides a menu allowing you to select the type of attack to carry out.

https://imghostr.com/fe7708_qwc

Let us discuss some features in SET.

- ❖ Infectious Media Generator: We use this feature to create a Metasploit payload with autorun features on USB or DVD device.

- ❖ SMS Spoofing: We use this feature to spoof specifically created SMSs to the target using a fake sender.

- ❖ Web Attack Vectors: We use these to perform phishing attacks on a target—mostly carried out using a fake link.

- ❖ HID vectors: This feature is the use of a programmable microcontroller that acts as a keyboard once plugged in. They include USB Rubber Ducky And Teensy USB HID both of which you can download from the resource pages below:

https://shop.hak5.org/products/usb-rubber-ducky-deluxe

https://www.pjrc.com/store/

For more information on SET, visit the resource page below:

https://github.com/trustedsec/social-engineer-toolkit

U3-pwn

U3pwn is a tool used to inject automated executable files into SanDisk smart USB devices. It contains the Metasploit Payloads injection package for SanDisk devices. It performs this action by removing the current ISO file and replacing it with a custom one with autorun capabilities. You can launch it by opening the terminal and typing u3pwn. This tool can fall under social engineering or gaining access.

Ghost-phisher

Ghost phisher is a python-based tool used for wireless audit and social engineering. It supports features that allow you to create fake networks that allow you to capture traffic on the network. It includes the following features:

- HTTP server
- Inbuilt DNS server
- Session Hijacking
- Web hosting
- Credential Harvester
- Metasploit Pentest Binding
- SQLite database support

- ❖ Fake Wi-Fi access point
- ❖ Inbuilt DHCP server
- ❖ Standalone update support

Ghost phisher provides a graphical interface using the python Qt Graphical library. Once you launch it, you just type the URL and perform the tasks needed.

Learn more about Ghost phisher from this GitHub resource page:

https://github.com/savio-code/ghost-phisher

Action step

Practice using the various tools discussed here. Other parts of this guide shall have practical practices that will test your ability to use some of the Kali Linux tools whose usage of we are illustrating in this guide. The more you practice, the easier and more intuitive using Kali Linux becomes, and the better a hacker/cyber security analyst you become.

In the next part of the guide, we move cover the reverse engineering tools in Kali Linux:

Part 13: Reverse Engineering Tools

Reverse engineering is the process of taking apart an already completed system or software and learning how it works backwards. In earlier days, the reverse engineering process only applied to hardware items but that has largely shifted over the years.

In this section, we are going to look at some of the tools within Kali Linux that help us take apart software so that we can learn how it works and the base infrastructure behind its operation.

Dex2jar

Dex2jar is a lightweight tool used to read Dalvik Executable files. It helps in disassembling of Android applications to jar files. The tool works with java and Android application classes. It comprises of several components that allow it to perform operations such as:

- Reading dex instructions to dex-ir format
- Converting dex-ir to asm format
- Representing dex instructions
- Reading Dalvik Executable files

❖ Disassembling and assembling dex file into and from smali files

For more information, peruse the Dex2jar Wiki guide on the resource page below:

https://sourceforge.net/p/dex2jar/wiki/UserGuide/

Java Snoop

JavaSnoop is a penetration-testing tool used to test the security of Java applications. It allows attaching of processes such as debugger, thus modifying method and injecting custom code. You can learn more about this tool from the following resource page:

https://code.google.com/archive/p/javasnoop/

OllyDbg

OllyDbg is an assembly level x86 debugger and analyzer for Windows applications. It is a very popular tool used to produce cracks for commercial software applications. It accepts an Exe or a dll file that it then disassembles and displays in binary thus allowing for modifications as illustrated below.

https://imghostr.com/ef15eb_vuk

This tool offers exclusive features such as:

- Direct disassemble of exe and dll files
- Interactive graphical interface
- Multithreading application support
- Searches whole allocated memory
- Dynamic frame stacks tracing
- UNICODE support
- Automatic recognition of ASCII & UNICODE strings
- Allows attaching of running processes
- Supports Borland format
- Portable
- Supports Object file scanning
- Dynamic code analysis – API calls, loops, strings, consts, etc.

EDB-Debugger

Edb-debugger is a cross platform AArch32/x86/x86-64 debugger. Its primary focus is modularity in reverse engineering. It offers features such as:

- ❖ Dump memory
- ❖ Conditional breakpoints
- ❖ Dynamic address inspection

Apktool

Apktool is a popular and effective tool for reverse engineering android applications. It has the capability of decoding and rebuilding entire application resources into native format. It offers step-by-step code edit with project-like structure.

Action Step

Practice using the various tools discussed here. Other parts of this guide shall have practical practices that will test your ability to use some of the Kali Linux tools whose usage of we are illustrating in this guide. The more you practice, the easier and more intuitive using Kali Linux becomes, and the better a hacker/cyber security analyst you become.

In the next part of this Kali Linux tutorial, we shall look at hardware hacking and the tools within Kali Linux for this task:

Part 14: Hardware Hacking

Hardware hacking is a very important aspect in penetration testing. It involves programming hardware to perform tasks automatically.

NOTE: Kali Linux does not have a wide serving of tools in this category:

The most important hardware hacking tool within Kali is Arduino IDE:

Arduino IDE

Arduino is a single-board programmable microcontroller. It comprises of a microcontroller (circuit board) and a programming environment used to write the code and upload it on the board. Arduino IDE uses simplified C++, which makes it easy to work with and learn.

You can find out more about Arduino and Arduino programming from the following resource page:

https://www.arduino.cc/

Part 15: Reporting Tools

Reporting is the last step after a successful penetration testing. It provides detailed information about the entire penetration-testing process/project. A penetration test report contains information such as methodologies used, results, and recommended security measures. Many penetration testers do not spend a lot of time on it because it is non-technical and involves a lot of writing.

In this section, we are going to look at some of the tools available in Kali Linux for reporting tasks.

Dradis

Dradis or Dradis framework is a penetration-testing tool written in Ruby to help cyber security professionals handle pen-testing reports easily. It helps organize all the penetration testing process, methodologies, and discoveries in one place. It also allows exports of saved information from other tools such as Nmap, Nessus, Burp Suite, etc.

How to work with Dradis

Here is how to work with Dradis

Step 1: Open the terminal and enter dradis command. Wait until it activates:

https://imghostr.com/3fa0a2_ja7

Step 2: Wait until it opens in browser. In case it does not, open the browser and go to http://localhost:3000

Step 3: Next, set the Dradis server password to use the app.

Step 4: Log in the credentials set above.

Step 5: On the left sidebar, you will notice a list of operations that will help in penetration.

Step 6: On the **All issues** tab, contains ways to add results in the library.

Step 7: On the **Methodologies** tab, it allows you to add the penetration-testing methodologies used. They include National Institute of Standards and Technology, Penetration Testing Execution Standards, Open Source Security Testing Methodology Manual, and Information Systems Security Assessment Framework.

Step 8: Next, continue by adding nodes and adding sub-nodes. Adding notes on each node and corresponding sub nodes.

After successful report collection and notes and screenshots addition, you can export the report into html formats.

Exportation of reports in pdf or word (doc, docx) version is not available in the community version of Dradis.

You can learn more about Dradis from the following resource

https://dradisframework.com/

Magic Tree

Magic tree is a reporting and data management tool for penetration testers. It provides an easy and straightforward interface for data consolidation, querying, and report generation. It also uses a tree-like project directory structure to organize files. It comes bundled in Kali Linux and is located under Reporting tools in the applications menu. Once you launch it, you can start by upload scan result from Nmap.

You can learn more about Magic tree from the Official web:

https://www.gremwell.com/

The next part of the guide covers Denial of Service Attacks:

Part 16: Denial of Service Attacks

Denial of service or DOS is a type of attack used by hackers and penetration testers to disable access or usage to an important service or resource causing inconvenience to the users.

DOS attacks do not collect any critical information. However, they combine well with other type of attacks making them deadlier. For example, a DOS attack on a server can cause the administrators to perform maintenance measures leading to a disabling of all security measures. This gives an attacker a chance to compromise the system and perform other operations such as uploading a rootkit or other malwares.

Understanding DOS

Denial of Service are unpredictable and do not have fast and hard solutions to prevent them. However, you can be able to detect them and perform necessary measures to disable them.

Denial of service attacks are mainly of three types:

- ❖ **Volume based attacks:** These are mainly the simplest type of DOS attacks. They occur when an attacker sends large volumes of packets to the target system using up all the bandwidth, which may result in system failure. The most common are UDP and ICMP floods.

- **Protocol Based:** These are the second type of dos attacks that work by using up all the target resources apart from the bandwidth. These types of attacks mainly focus on resources such as firewall, intrusion detection systems, and network switches. They include attacks such as Smurf attacks, Fraggle attacks, SYN floods, and many others.

- **Application layer attacks:** Application attacks are deadlier as they attack the application layer itself. They mainly focus on specific vulnerabilities and comprise of a fake application layer. They involve sending application layer requests with the intent of crashing it. Their main targets are usually Apache servers and Microsoft IIS.

Tools used in Denial of Service Attacks

We have more than a thousand tools that you can use to perform performing denial of service attacks. In Kali Linux, we are going to use Metasploit auxiliaries to perfom DOS attacks.

Start by opening the terminal and then navigate to the following directory /usr/share/metasploit-framework/modules/auxiliary/dos

This lists all the type of dos auxiliaries available as shown below:

https://imghostr.com/b0291f_8vo

Once in the directory, you can navigate according to the type of DOS to perform such as windows, apple, android and such.

Xerxes

Xerxes is very powerful denial of service and stress-testing tool. It has the following features that allow it to work very efficiently for denial of service:

- ❖ Multiple Attack Vector
- ❖ TLS support

- ❖ HTTP header randomization
- ❖ Multiprocessing
- ❖ User agent randomization

To use it in Kali, open the terminal and clone the following repository:

https://github.com/sepehrdaddev/Xerxes.git

Navigate to the cloned directory and execute it by typing ./Xerxes -h to show list of commands required.

Action step

Having covered all the major features and tools provided by Kali Linux, you now have a fairer idea of how to use the Debian-based Linux distribution to perform for penetration testing and cyber forensics.

As mentioned severally, the best way to get better at using these primary tools is to practice using them as often as possible until you truly master Kali Linux. To help you practice what you have learned, the next part of the book gives you a hands-on project that you can use to exercise your command and understanding of the various Kali Linux, cyber security, and hacking tools, features and strategies we have discussed in this Kali Linux masterclass.

Part 17: Hands-on Kali Linux Projects: Exploiting Metasploitable 2

In this section, we are going to perform a comprehensive attack against Metasploitable 2 using the techniques learned throughout this guide. We will start with network scanning and then move on to exploiting any vulnerabilities found.

Here are the services we are going to exploit in the Metasploitable machine:

- ❖ Network Scanning using Nmap
- ❖ Exploiting FTP
- ❖ Exploiting SSH
- ❖ SSH RSA exploit
- ❖ Telnet exploit – brute-force and credential capture
- ❖ Java Exploit
- ❖ VNC exploit
- ❖ PHP CGI (port 80)

Requirements

In this section, we are going to require the following utilities and tools:

- ❖ Kali Linux – attacker
- ❖ Metasploitable 2 – Target system

Network Scanning – Nmap

The first step we are going to perform is to scan the target machine for running services. We are going to achieve this using Nmap (terminal version)

Open the terminal and enter the nmap -sV -p- <ip address >

This will scan the Metasploitable machine for all running services and their versions.

https://imghostr.com/7a6a36_vt5

From the above scan, we can see some services that run of the Metasploitable machine such as VSFTPD version 2.3.4, OpenSSH, VNC and more.

Nmap gives detailed information about specific services. Once you have the information, you can use it to find its vulnerabilities either using searchsploit or exploid-db website.

Exploiting FTP

Let us try to crack the ftp password using Hydra.

The first set is to create a username.txt file and save it to the desktop, in the file include the following usernames:

https://imghostr.com/755013_srr

Next, create a passwords.txt file, save it to the desktop, and include the following passwords:

https://imghostr.com/10a3f9_5ch

Open the terminal and enter the command hydra -L ~/Desktop/usernames.txt -P ~/Desktop/passwords.txt <ip_of_Metasploitable> ftp

https://imghostr.com/2e9651_b2e

Hydra will start the cracking process using the list of usernames and passwords contained in the wordlist. From the above screenshot, we can see that the tool found 17 passwords. In Metasploitable 2, it is acceptable to use ftp (lowercase or uppercase) as username and any other password.

To test whether the credentials are correct, let us try to connect to the ftp of the Metasploitable 2 machine.

Step 1: Open the terminal and install ftp using apt-get install ftp.

https://imghostr.com/e30d7c_29d

Once installed, type ftp <ip address_of Metasploitable> and enter either one of the usernames and passwords cracked from above.

https://imghostr.com/f097bd_c9y

This allows you to connect to the Metasploitable machine via ftp to upload and download files to and from the computer.

Exploiting SSH

In this section, we are going to exploit the SSH service using an auxiliary and upgrading the session to Metasploit Meterpreter. The first step is to search for the auxiliary using the search command in Metasploit.

Next, use the module and set rhosts to the Metasploitable machine. After that, set the user_file as the usernames we created earlier (provide the full path of the file), and then set the pass_file as the password file (created above). Once done with this, set the stop_on_success as true to allow the auxiliary to stop scanning once it finds the correct credentials.

Lastly, exploit the module:

https://imghostr.com/76d9oc_fqo

If the process finds the username and password, it will open a command shell session. To elevate the session to a Meterpreter session, use the command **session -u <session_number>**. Then use the **session <session_number>** to interact with the Meterpreter session.

SSH RSA Exploitation

Here, we are going to brute-force the SSH RSA key using a python module built in Metasploit:

For this process, the first step is to download the RSA files available from the following resource page:

https://github.com/offensive-security/exploitdb-bin-sploits/raw/master/bin-sploits/5622.tar.bz2

Once downloaded, extract the file and copy the entire directory into /usr/share/exploitdb/exploits/linux/remote/

Open the terminal and navigate to the above directory. Lastly, enter the command python 5720.py rsa/2048/ <ip_of_metasploitable> root

https://imghostr.com/55f133_os7

Wait until discover of the correct RSA key occurs. It provides the command to execute to connect to SSH of the target machine:

https://imghostr.com/a36756_buj

Once connected, you get root access to the Metasploitable machine and can execute commands.

https://imghostr.com/2799f9_tn7

Telnet Exploitation

In this part, we are going to see how to use Wireshark to sniff passwords using telnet in Metasploitable 2.

First, open Wireshark and select your network interface to start scanning. Next, open the terminal and enter command telnet <ip_of_metasploitable>

https://imghostr.com/3802f7_dk2

Go back to Wireshark and filter the collected traffic using tcp.stream. Under Wireshark menu, go to Analyze – Follow Stream and select TCP Stream.

https://imghostr.com/ef989c_qwa

To exploit Telnet using Metasploit auxiliary module, open Metasploit and search telnet_login. Next, set the rhost to Metasploitable ip, user_file to the username file and pass_file as the password file (previously created above)

https://imghostr.com/7beda6_jyt

Launch the exploit to start the brute-forcing process. Once the auxiliary module finds the correct password, it will open a session that you can then elevate to a Meterpreter session.

https://imghostr.com/ad339d_q02

Java Exploitation

We are going to gain access by exploiting the default configuration in the Java RMI registry that allows loading of classes from a remote URL using the Remote Method Invocation exploit running on port 8080. It calls the RMI Distributed Garbage Collector method available in every endpoint making it vulnerable. Open Metasploit, search for the rmi exploit, and then set the required arguments and exploit:

https://imghostr.com/a7eed8_516

Virtual Network Computing

We can exploit the VNC service running on port 5900 (according to previous Nmap scan). Using the vnc_login module that finds login credentials, we can be able to connect to Metasploitable on VNC. Open Metasploit and search vnc_login auxiliary. Once it finds the credentials, you can launch the vnc viewer and connect.

https://imghostr.com/0a5121_nga

NOTE: In the above illustration, the VNC window appears in the background

PHP (CGI)

Metasploitable 2 runs PHP as its backend server language. If we can manage to find the php configuration info, we can find out whether it is running as a CGI. Open the browser and open <ip_of_metasploitable/phpinfo.php> to display php information.

https://imghostr.com/fe9ed0_knu

On PHP 5, running PHP as CGI opens the vulnerability of argument injection that you can then exploit using -d flag in php.ini. Open the Metasploit and search php_arg exploit. Set the required parameters and exploit.

https://imghostr.com/e8ad9f_huw

Once exploited, it opens a Meterpreter session allowing for interactive control of the system.

Action step

Run through the various hands-on tutorials provided in this part of the guide several times until you fully internalize the working principles behind using the various features and tools within Kali Linux.

Keep in mind that the more you interact with these tools through practice, the more intuitive their use shall become to you, and the better you shall become at using Kali Linux to hack systems and perform penetrating testing and cyber other forensics and security tasks.

Conclusion

As we have mentioned severally, Kali Linux has hundreds of tools. In this tutorial, what we did was cover the primary ones that relate to hacking and cybersecurity basics.

Use what you have learned from this guidebook as a springboard that allows you gain hands-on experience with using Kali Linux comfortably. Once you master how to use the various Kali Linux tools and features discussed in this guide, you will be well on your way to mastering Kali Linux.

Thank you for reading this ultimate Kali Linux guide for beginners and intermediates.

If you are happy with what you've learned, don't forget to leave a review of this book on Amazon.